BACK FROM THE BRINK
AMPHIBIANS

BLACK RABBIT BOOKS | JOANNE MATTERN

TABLE OF CONTENTS

Contents

Chiricahua
Leopard
Frog

Do you hear snoring? Nope! That is a Chiricahua leopard frog making noise.

These frogs live in Arizona and New Mexico. They need water to lay their eggs. However, a lot of water has disappeared. Ranchers drained ponds so they can raise cattle. By the late 1990s, there were only 80 places left for the frogs to live.

People worked together to save the frogs. They made new water tanks. Zoos bred the frogs. Then they released the frogs. Today, there are more than 10,000 of these frogs in the wild.

Did You Know?
These frogs also face trouble from
bullfrogs that eat their eggs.

2

6

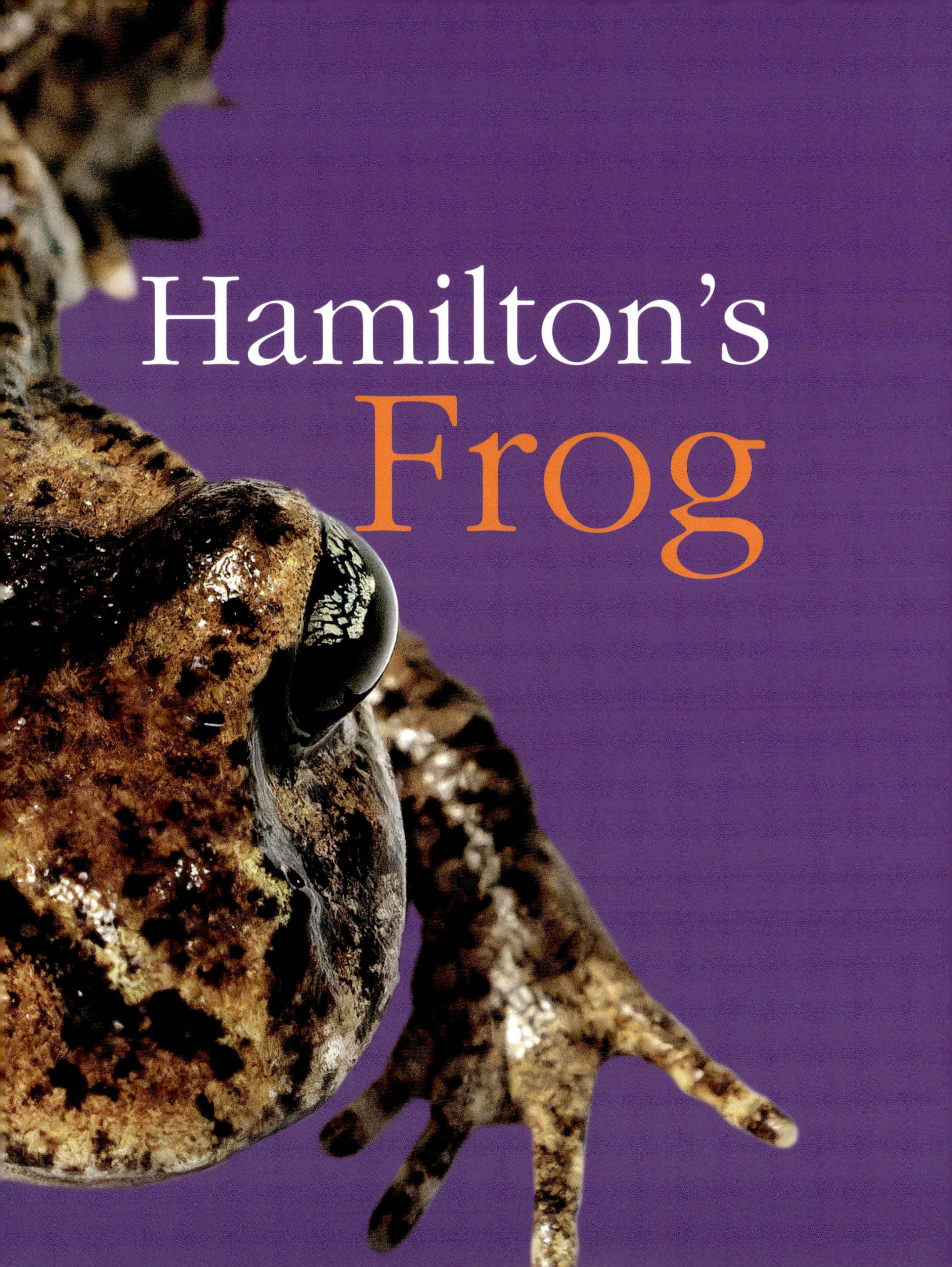

Hamilton's
Frog

The Hamilton's frog is tiny. It is less than 2 inches (5 centimeters) long. These frogs once lived all over New Zealand. However, many predators killed the frogs. By 2004, there were only about 300 left.

Scientists worked hard to save these tiny amphibians. They created a new place for the frogs to live. Then they took 80 frogs to their new home. Today, many frogs live in two areas in New Zealand.

Think About It
Why do you think predators are more
of a threat to one species than others?

Hellbender
Salamander
3
10

Hellbender salamanders live in fast-moving streams in the eastern United States. They need water to survive. But dams were built in their homes. People polluted the water. By 2011, the animals were almost gone.

Then people stepped in to help. They cleaned the rivers and streams. They made the rivers healthy places for the salamanders to live.

Zoos helped too. They raised some of the salamanders in captivity. Then they released the animals back into the wild.

Think About It
Amphibians can be a clue if water is clean. Why do you think that is?

12

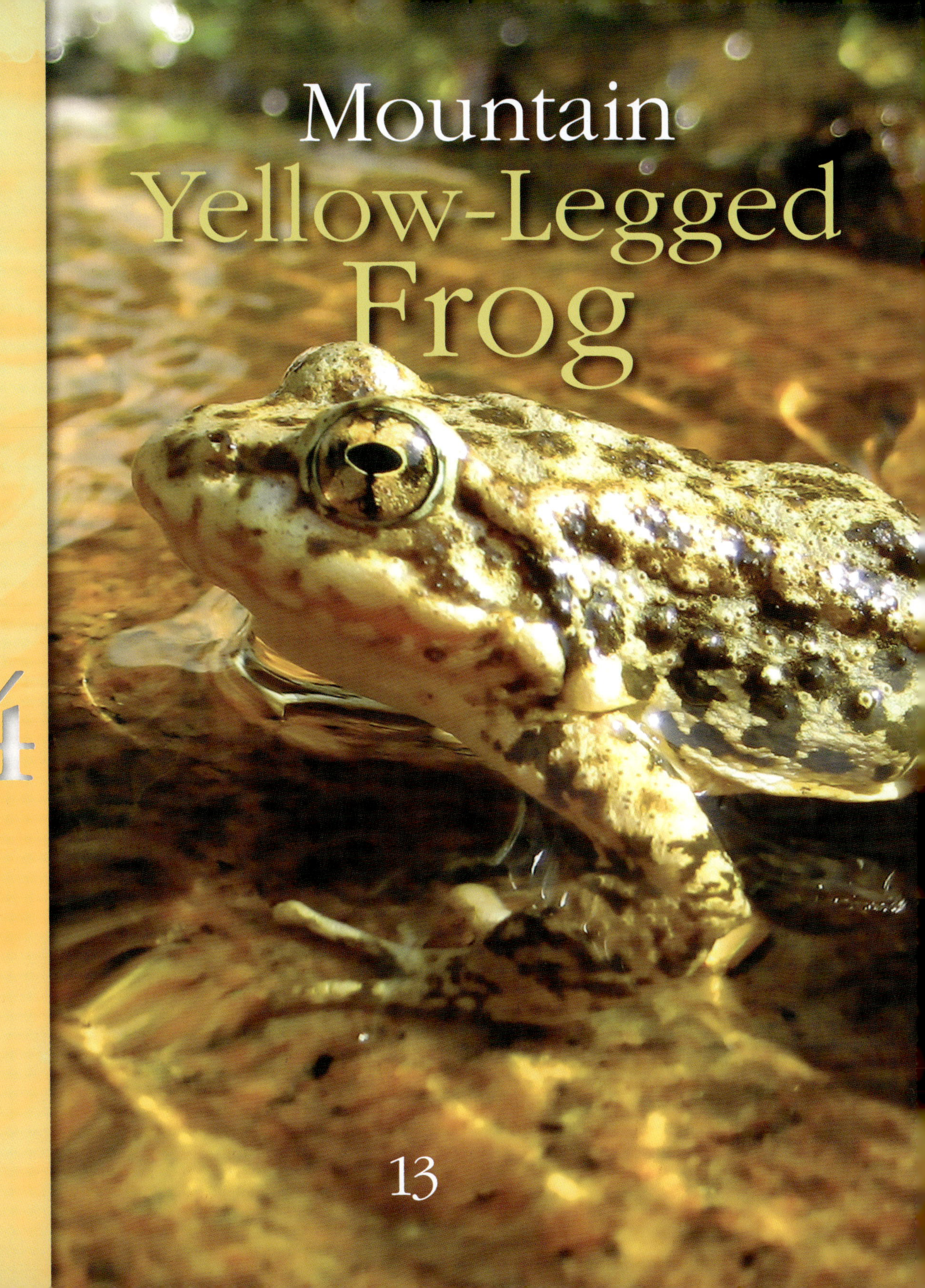
Mountain
Yellow-Legged
Frog
4
13

Most frogs live where it is warm. But not the mountain yellow-legged frog! It lives high in the cold mountains of California.

These frogs were once common. Then people wanted to fish for trout. Many fish were added to the frogs' lakes. The trout ate the frogs' eggs. Soon there were hardly any frogs left.

People stepped in to save the frogs. The government removed the trout. Zoos raised the frogs. Then they put the frogs back into lakes. Today, these frogs are doing well.

Did You Know?
These frogs hibernate at the bottom
of lakes during the winter.

15

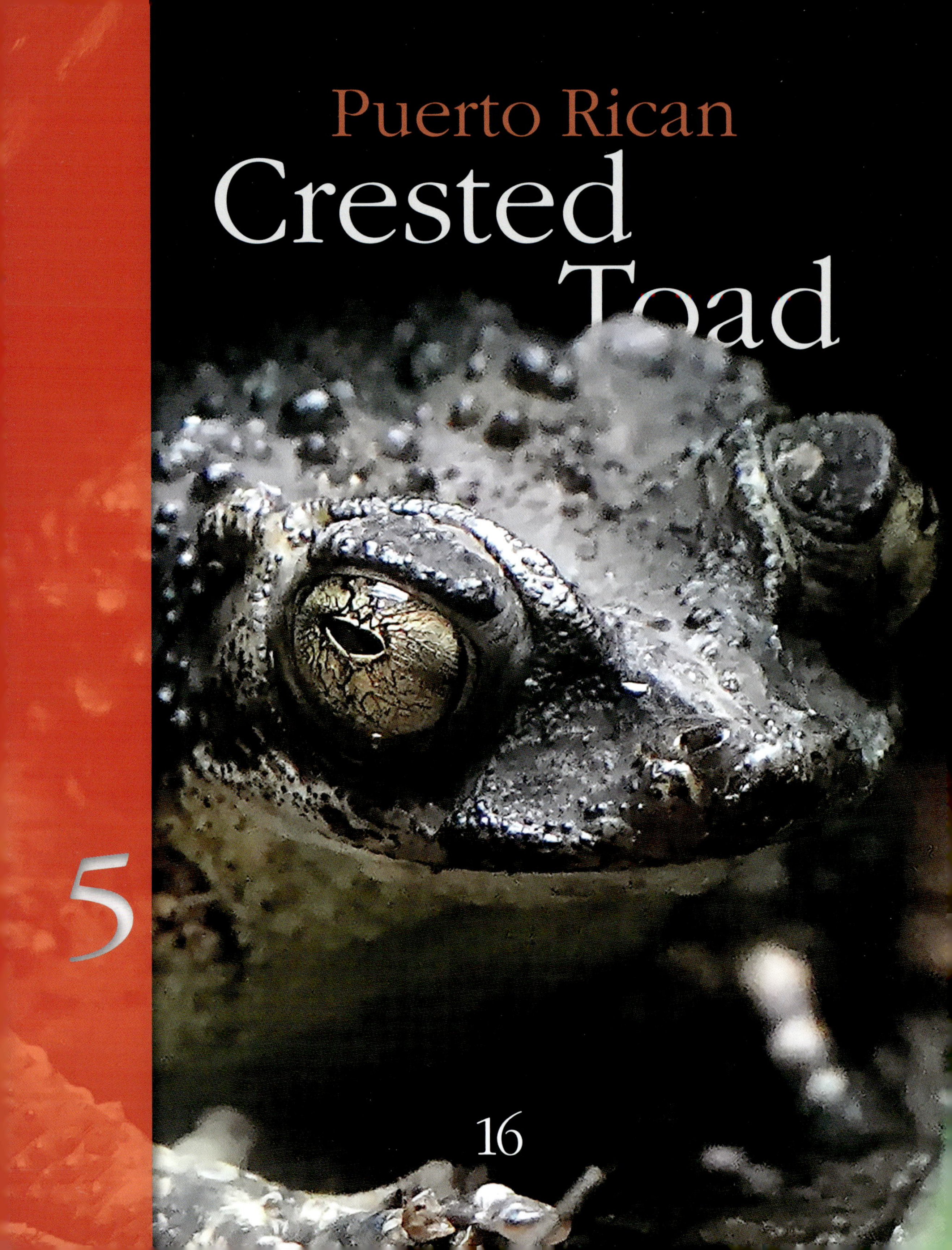

Puerto Rican
Crested
Toad
5
16

Scientists first saw the Puerto Rican crested toad in the 1860s. Then the animal disappeared. For years, people thought it had died out. Then a few were spotted in the wild. Scientists had a chance to save the toad.

Scientists captured some of the toads. They sent them to zoos in the United States and Canada. The toads were bred there. Then the zoos released thousands of tadpoles into the wild. This toad is still in danger. But now it has a chance to survive.

Think About It
Why are zoos important to
saving endangered species?

18

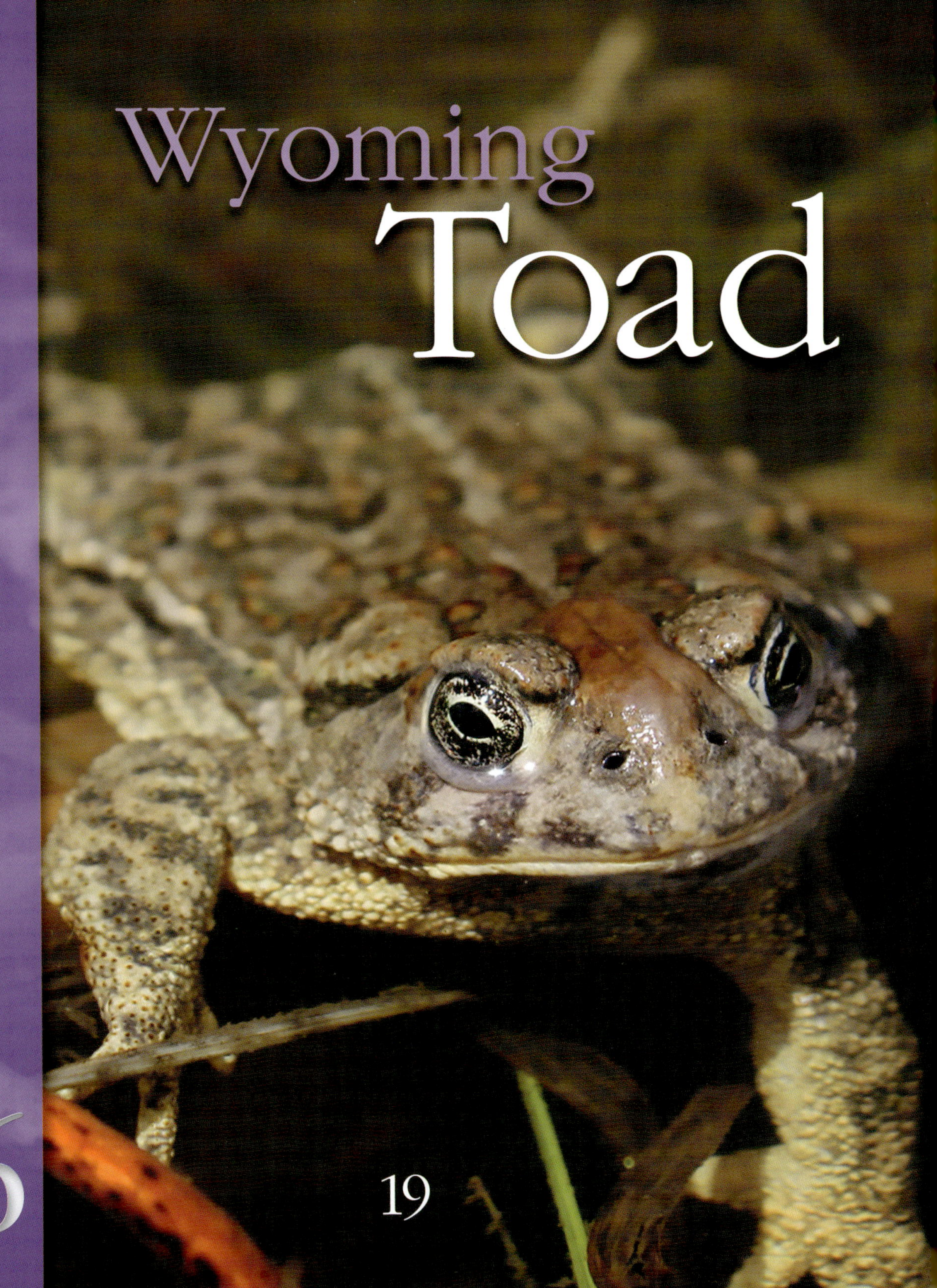

Wyoming Toad

19

The Wyoming toad only lives in one place in the world. But disease and **habitat** loss put this toad in danger. Scientists were very worried. In 1994, they captured the last 10 toads.

The toads were raised in zoos. Then, in 2023, they got a new home. The government created a **refuge** for them. It is near Laramie, Wyoming. Now these toads have a safe place to live.

Did You Know?
The refuge is the only place Wyoming toads live in the wild.

MORE TO EXPLORE
WHERE IN THE WORLD?

MOUNTAIN
YELLOW-LEGGED
FROG
California

CHIRICAHUA
LEOPARD FROG
Arizona and
New Mexico

WYOMING TOAD
Southern Wyoming

PUERTO RICAN
CRESTED TOAD
Puerto Rico

HELLBENDER
SALAMANDER
Eastern United States

HAMILTON'S FROG
New Zealand

FANTASTIC FACTS

Wildfires often destroy the long grass where the Chiricahua leopard frogs nest.

Hamilton's frogs do not have ears or vocal cords.

Hellbender salamanders are covered with slime. They look poisonous, but they aren't.

Mountain yellow-legged frogs lay on sunny rocks to keep warm.

The Puerto Rican crested toad is the only native toad in Puerto Rico.

A Wyoming toad can release poison from the top of its head.

COOL COMPARISONS

These animals are making a comeback.
How many live in the wild?

**Chiricahua
Leopard Frog**
about 10,000

Hamilton's Frog
about 300

**Hellbender
Salamander**
about 915

**Mountain
Yellow-Legged
Frog**
about 200

**Puerto Rican
Crested Toad**
between 1,000
and 3,000

Wyoming Toad
about 1,500

RESOURCES

Glossary

amphibian (am-FIB-ee-uhn) A cold-blooded animal that can live on land and in water.

captivity (kap-TIV-i-tee) The state of being kept in a place, such as a zoo.

endangered (en-DAYN-jurd) Close to no longer existing.

habitat (HAB-uh-tat) The place where a plant or animal grows or lives.

hibernate (HI-bur-nayt) To spend the winter sleeping or resting.

predator (PRED-uh-tuhr) An animal that eats other animals.

refuge (REF-yooj) A place that provides shelter or protection.

Read More

Hughes, Sloane. *20 Things You Didn't Know about Amphibian Adaptations*
New York: PowerKids Press, 2023.

Thomson, Sarah. *Save the Frogs.* New York: Philomel Books, 2023.

Index

TOP RANK is published by Black Rabbit Books, P.O. Box 227, Mankato, MN, 56002. • COPYRIGHT © 2026 Black Rabbit Books. All rights reserved. No part of this book may be reproduced in any form without written permission from the publisher. • Top Rank is an imprint of Black Rabbit Books • Series designed by Danny Nanos • Book designed by Jason Knudson • Photographs © Alamy Stock Photo/John Cancalosi, 5, 21, 23, Kevin Schafer, cover, 1, Natural History Collection, 15, Tom Grundy, 14; © Andy MacDonald/CC BY-NC 4.0, 9, 21; Flickr/Bill Dorsey, 16, Charles (Chuck) Peterson, 4, Shannon Buttimer, 20; Minden Pictures/Joel Sartore/Photo Ark, 2, Pete Oxford, 10–11; Oscar Thomas, 6–7; Shutterstock/Nastya Tsvyk, cover, 1; USGS/Public Domain, 13, 21, 23; Wikimedia Commons/brian.gratwicke, 12, 21, 23, Jan P. Zegarra, 18, U.S. Fish and Wildlife Service Southeast Region, 17, 21, 23, USDA Forest Service, 19, 21, 23 • Printed in India
Library of Congress Cataloging-in-Publication Data: Names: Mattern, Joanne, 1963- author. I Title: Amphibians: back from the brink / by Joanne Mattern. I Description: Mankato, MN : Top Rank is an imprint Black Rabbit Books, [2026] I Series: Back from the brink I Includes bibliographical references and index. I Ages 8–11 I Grades 2–3 I Identifiers: LCCN 2024043348 I ISBN 9781644667965 (library binding) I ISBN 9781644668283 (paperback) I ISBN 9781644668603 (ebook) I Subjects: LCSH: Rare amphibians—Juvenile literature. I Amphibians—Conservation—Juvenile literature. I Endangered species—Juvenile literature. I Wildlife recovery—Juvenile literature. I Wildlife conservation—Juvenile literature. I Classification: LCC QL644.7 .M373 2026 I DDC 597.8—dc23/eng/20250110 I LC record available at https://lccn.loc.gov/2024043348